# The Second Consciousness
Of the complexity of life

Christian Hermenau

# Contents

1. Number logic — 7
2. What is behind the quanta — 13
3. The magic of food intake — 19
4. Heisenberg and wheat grains — 25
5. The search for the magic — 29
6. Mathematics and quantum mechanical particles — 35
7. Computers, switches and transistors — 43
8. The rhythm of the atoms — 49
9. The entropy in the networks — 61
10. A diversified world — 71
11. The first consciousness — 79
12. Entanglement — 83

| | |
|---|---|
| 13 Black Holes or not | 91 |
| 14 something million light years away | 99 |
| 15 Live stream between the particle | 103 |
| 16 Quantum computer | 109 |
| 17 Quantum computers and networks of life | 113 |
| 18 Reference/Index | 117 |

# 1 Number logic

It is not easy to come into this world at the beginning of a life and it is not easy to leave it at the end. But most of our life time we go uninspired through the everyday life and hope for a miracle, although everything around us is magical and we ourselves are the biggest wonder in it.

We can also trivialize things on earth in such a way that we only see the simple things behind it, the laws behind everything, which all bodies and beings have in common. Thereby we abstract the world. We then describe the whole as idealized and absolute. It is fascinating that $1 + 1$ in it are always 2. We create a number system that is boundless, but also functions according to very simple laws. We could count a whole life long and would only perhaps reach the billion and there would be no end in sight. The numbers go on and on, with this simple dull logic. There would never be a surprise, never would we come across a number that is wrong or doesn't fit. This is the cool abstraction of numbers. Something man-made that can be applied to grains, to cars, to stars or people. One hundred thousand grains of sand are about 20 grams. We could count them lifelong and as a number they would be absolute, but two hundred thousand grains counted are not equal in weight. Above a certain accuracy, differences become apparent, at the latest in the microgram range. We could count a lifetime of one hundred thousand grains of sand and would always weigh different masses, because no two grains are the same, only the number is the same.

So, what is wrong if we transfer pure mathematics, a pure, simple, exact order, together with its logically conceived laws, to the physical world and the result is not absolute and one hundred percent, but each Something has its own reality. Every grain of sand is unique, but our formulas of reality do not allow this uniqueness. Physical formulas should ideally function exactly like the mathematically abstract laws from which they are constructed. Nature is geometrized in them and written down, for example, in field equations, which then describe an infinitely fine space in an infinitely fine time - all in theory, in the way we thought it up. And then there are atoms, small packets of energy that are at the same time punctiform and yet not infinitely small, that have an expansion like a long wave and are then supposed to be compact like strings again, but are mathematical and thus remain abstract. On the one hand, it is possible to build atomic clocks with the timing accuracy of atoms, in which two such clocks differ from each other by only one second in a time longer than the age of the universe. On the other hand, such a thought, abstract, elementary particle is neither exactly determinable nor exactly positioned. From our point of view, it is blurred.

How can we believe that atoms function like numbers, that space, time and matter are exact, when in our world nothing is really one like the other? Not like the numbers. There are no two particles that are the same in everything, unlike our numbers. Although 231 is not 232, but except that one number is by 1 greater than the other there is no difference, nothing surprising, nothing individual or special. If we transfer numbers to life, then 231 flies are not equal to 232 fleas. Then not even 231 grains of wheat are equal to 231 grains of wheat, not if we look closely at each grain.

We take the essence, the soul, out of a grain of wheat and all that remains is a number. For these numbers there are special connections and laws, but do these laws then also apply to the living grain? Can something big, something built up work in the same way as something idealized, something imagined? One could go even further into the smallest details here, because even the most elementary building blocks are not all the same. Each has its own state, its own network of individual energies stored. No particle is like another. Although one might think that the elementary building blocks would have to be perfectly idealized and thus would fit exactly to a cold abstraction, this is not the case at all. We transfer a macroscopic world to the elementary realm in the belief that we will encounter the small, indivisible, divinely perfect spheres of the ancient Greeks. The large complex world seems to fit the abstractions in an idealized way, but the very real conditions that make the movements so confusing do not disappear when we look at the exact basic building blocks, the atoms. All of a sudden, the atoms are not so exact anymore, they are more confusing and complex and so completely different from our familiar living world. Suddenly it becomes apparent that with only 53 qubits, only 53 real building blocks that have no contact with our world for just three minutes, it is possible to calculate mathematical tasks that would take conventional supercomputers many thousands of years to complete. 53 systematically networked atoms surpass every supercomputer - only 53!

But our world consists of an unimaginable number of particles, which may be all are networked, or put another way, why should they not be networked? And if they are, what can be calculated with them or is it the very secret of consciousness? If we think about the world, about this universe, its size, its abundance, or if we take a closer look at our earth, the life on it, a queasy feeling arises. We then feel very clearly that there is something incomprehensible, something beyond comprehension, something that exists beyond all knowledge and rationality. The whole thing is so big, so much bigger than we can think and yet we are a part of it in the middle of it. No matter whether we are believers or scientific thinkers, when it comes to the questions: where does everything come from, why do we exist, why are we so complicatedly constructed, it makes us shiver. For the most part of our time we don't deal with such questions, maybe some of us ignore them completely and live their lives, but we would all like to know the answer to them, all of us: gods as well as humans, every thinking, whether it is here or somewhere in the universe. We realize that an answer is beyond our comprehension, and probably no almighty God can understand it either, because it is more incomprehensible than omnipotence.

Although we may be able to understand the structure of our universe, to find a self-contained connection, perhaps to discover the mathematical laws according to which everything functions here, if it is possible according to laws, but with that we have not answered the overriding question of why all this, the where from and where to.

# 2 What is behind the quanta

If we can't just come to life, then we can't just die.

If it is not just the gene code in which all our characteristics are stored that regulates everything in the germ cells, if it does not just run like clockwork when new life is created, then what is it? What makes one cell into billions of cells by continuous division, which are not just simply there, but ordered to a maximum degree? We are not thinking of a precise order as in a diamond or as in an ice crystal, but each cell is individual and yet connected to the whole. No matter whether microorganism, plant or animal, the cells that are connected to each other are from the very beginning livingly networked, work-sharing and differentiated. Also, life was already there before, the liveliness does not only arise during the division. It forms a whole without being hierarchical. It exists with and without consciousness, with and without feelings, depending on what is being created. It is always the same starting parts: Tissue cells that divide. Whether it then becomes a worm or a human makes no difference with the initial cells. The principle is the same. Worm cells do not differ from human cells in the beginning. The genes also always consist of the same base pairs, only arranged differently. And although we are able to observe and film cell division and differentiation today, although it seems to run like clockwork, there must be a lot more behind it. A much greater knowledge, a much, much greater network, in which not only the arrangement of the cells is stored, but also the experience that millions of beings have had before: So not only the arrangement of the atoms in space, but also how the arrangement has changed and developed over time. Babies or microorganisms must have a stored world knowledge of the past in them from the beginning, the fly just like humans. And this knowledge is not only contained in the genes that run according to plan. This is too technical, too simply thought. If the highest, the most complex thing in this universe takes place on the gene molecule and cell level, then our higher complex life would indeed be so unimaginable, so fantastic, that we would always have to introduce a God who created us like a master builder and then breathes the soul into us with his divine breath. We need another foundation, another basis on which all this is created. And it is supposed to have happened in this universe, with the means that are available to us here, out of itself. We do not want to strive for infinity. Not, for example, to introduce infinitely many universes, which all exist parallel to each other and even influence each other somehow. We do not want that! We want it all to happen in our one universe only. In the meantime, people have reached a knowledge of nature that shows us that this is actually possible with the materials we find here in the universe. It is possible to explain life with atoms as we find them here in the universe. We only have to allow some changes and change the centre of gravity, then life could be explained with what we have. Genes are made up of atoms and atoms are subject to quantum mechanics, but behind quantum mechanics are the networks that make the connection to abstract consciousness.

Einstein and Bohr discussed long and heatedly at that time whether there is a possibility to observe the elementary particles without measuring them. Einstein was firmly convinced that, like a mouse in a corner, one can only observe the universe without influencing it in any way and relaxed one thought experiment after another. For Einstein a lot depended on it, after all it was about whether we could understand the universe in principle or whether there were areas to which we had no access. But Nils Bohr persistently disproved Einstein's ideas again and again and was later to be right. Thus, the Copenhagen interpretation of quanta prevailed, which assumed the impossibility of observation without measurement, thus granting undreamt of freedom to particles and photons.

For physicists, this also means that in a pair process, for example, the state of the two particles or photons remains undetermined as long as it is not measured. Or, which is the same thing, reacts with the outside world. This means that a quantum-mechanical state can be one, zero or zero/one superimposed and remains so as long as neither of the two objects reacts with the outside world, no matter how far away the particles should be from each other by now. Two important extraordinary properties of quantum mechanics come into play here: the superimposed, i.e., ambiguous state and the entanglement, the special contact between two particles or quanta that are not yet separated and yet are far away from each other. But we will go into this in more detail later.

For our living processes it is not important whether we have access to the particles, even if we can then possibly prove our hypothesis only indirectly. For us, the extremely high degree of interconnectedness of matter plays a very important role and we want to get to the bottom of it. According to our structure, elementary particles such as protons and electrons are not only in permanent contact with an endless number of other particles via gravity, but they also store each of these contacts additionally. This means that we have extremely large networks very quickly. With the numerical values and process times that play a role here, it is completely hopeless for us to be able to track anything for even one second. We can only calculate the statistical expected values of place and time or momentum and energy of the particles and then we are in the relatively rough area of quantum mechanics. For us macroscopic beings this represents the lowest limit. This does not mean that there is not much more behind it, but only that we are denied access to it. The objects, particles and quanta are small, but tangible and real. The networks above are not. They are fantastically big, incredibly fast and the storage capacities are beyond our imagination. But it remains the most difficult of all questions: can intelligence and consciousness also develop in such randomly created networks? Is there a second intelligence besides ours in this universe?

Only that matter is much networked does not necessarily mean anything. The many may create the conditions for consciousness, but there is no automatism that makes many things comes alive immediately. And yet the question remains: is there a way in which consciousness can also develop differently, without the central brain of the human being? And is it perhaps the other way round, that first this higher consciousness arose, which then involved in the development from inert matter to complex beings?

# 3 The magic of food intake

We kill animals for food or less dramatically, we feed on plants. In doing so, we assimilate the plant. For example, when we eat a carrot, we chew it up, salivate it and swallow it. In the stomach, digestive juices are added. The life of the carrot is destroyed. The carrot and with it all the atoms and molecules are enclosed by us, by our atoms. Then something is broken up that takes the life out of the carrot and then reassembles it in us. It is not the atoms that cannot be killed, nor the molecules. A self-contained living system is enclosed by another apparently far superior system and broken down into its components, in order to then use them as building material or for burning in one's own system. But who does this? Who in us knows how to make something for our own bodies from something living foreign? We do not know, and neither do the researchers know how to make something so complicated. How the body knows, or what in it knows how organic material is transformed into the body's own. Just as germ cells know exactly how to divide up to make a plant or an animal, so something on an elementary level knows what is needed where and how to decompose it. It is a knowledge that is not stored in the brain. This knowledge works just as well with the amoeba or the bacterium as with large plants and animals. We humans may be far superior to plants, but the elementary structure that regulates everything is the same. Complex plants have the same genes, the same gene code as we humans and animals, and the intake of food, the burning and incorporation of foreign material into the body does not suddenly change just because it happens in humans. Only, what or who regulates this elementary liveliness, who knows how it works?

The food pulp has to be analysed; something has to know what it is made of. It can't just be destroyed and then the atoms are put together properly. The infinite number of possible foreign proteins must be broken down into their elementary components again, but these are not the atoms or any molecules, but the eight essential basic amino acids and that is important. A large foreign protein must therefore be analysed, what do we have here, and then be broken down again at the right places. At the atomic level into a specific amino acid and this is then removed accordingly and brought to where it is needed. But who, whom or what is needed where, how do it know? Furthermore, we are not talking about a single protein that the whole body is exclusively concerned with all day long: it only analyses, thinks about what to do with it, asks around and then takes it to a suitable specific place. No, we are again dealing with the incomprehensible many things and the fact that everything seems to happen at the same time. Huge amounts of proteins have to be processed simultaneously. When we eat, then also not proteins, fats and carbohydrates nicely separated in large intervals one after the other, no, when, then everything comes together. Even the most beautiful gourmet food is chewed in the mouth and at the latest in the stomach it is properly mixed and pre-treated with digestive juices. All food taken separately is treated the same way there. So apart from the fact that we eat very consciously, that we look at it very carefully before we eat it, that we consciously know if we are hungry, what we are hungry for and today very selectively distinguish between disgusting and lustful food, the comminution process in the mouth and the kneading process together with the digestive juices in the stomach always proceed in a similar way. But then suddenly something magical happens, something hard to understand, something that extracts the essential ingredients from the unsightly stomach pulp and distributes them throughout the body in a targeted manner. As soon as the material is transported through the stomach and intestinal wall in the blood, nothing is unsightly and undefinedly mushy anymore, then everything has its own order. Now we find ourselves in the area of a closed body with its organs, muscles, tissue and ordered body cells, which is permeated by blood and body juices of the lymphatic system. But not disorderly, like an external sequence of events, like a cyclical ebb and flow of tide that washes over the mudflats, but in precisely regulated, finely branched paths to all muscles and body cells. Everything is supplied specifically and knows about each other. Whoever needs energy, where fat is stored and when something needs to be repaired or rebuilt. Who develops something like this, who regulates it, who has an overview of the whole? The brain does not. A plant, for example, a tree, despite its impressive size, does not have a central nervous system and yet it functions in exactly the same way as in us or higher animals do. Well, there is an army of microbes, the microbiome, without which digestion would not function. But then how do the microbes know what to do? Do they only specialize in certain foods and are constantly searching for them? Hardly, they know it and we do not know how they know it.

In the human digestive system, there are almost as many nerve cells as dogs or cats have in their central nervous system. Almost as if we have outsourced a second, much smaller brain. And apparently digestion plays a central role and is much more complicated than we thought, because this second brain is the tribal-historically much older one. It looks like digestion is meticulously monitored and is in constant communication with the cerebrum. Our well-being and emotions are determined by the stomach, not only metaphorically but also by a real complex nervous system around the digestive process. But even this outsourced brain only controls the external processes, it does not know itself how to sequence amino acids, convert sugar into fats or carbohydrates into sugar. Our body turns food into a mushy mash, but then it doesn't know what to do. Then beings come into play that know how to do it although they are so simple: Microbes and fungi.

Even though they are so simply constructed and we feel so indescribably superior, they know how to do it and we do not. But, does a thinking person need to know how a processor works or be able to control it just to use a smartphone? Of course not, at least not every single, not every simple user. He just has to learn to use the bits and bytes to his own advantage or disadvantage. If, at all, the network of humanity as a whole needs to know how to do it: How to build processors and how to make and program smartphones. We are many, we build on the knowledge of generations before us and we have specialized. Through this we have become immensely powerful and the conscious rulers of the world; At least on a macroscopic scale. Nevertheless there is a huge difference between bits and bytes and bacteria, viruses and fungi. Technical devices, processors, transistors are all not alive in contrast to our army of microorganisms. Our consciousness has to enter everything from technology into computers. Nothing arises out of itself, nothing unforeseen. The microbiome, on the other hand, has a life of its own, because they are all self-sufficient beings. Our abdominal nervous system or our cerebrum has not programmed or created anything.

The microbes have been around much longer than we have. Yes, the simply constructed microbes have been here on earth almost since the beginning of the world and know the secret of photosynthesis or the breakdown of macromolecules into meaningful units. They are much older, they seem to have stored the knowledge in themselves and there are many of them who seem to be able to exchange information and make arrangements with each other, however they do it. But they seem to be much more effective than we can imagine. There is a life of their own they lead and so they can even indirectly influence our feelings. So, we have to ask ourself who is dominating whom.

Maybe this symbiosis is much more, just like big networks as a whole are much more, but maybe they are still too little to explain this complex world. Because we still have no idea what life is all about, how something can become alive, when something is called alive and how far it goes.

# 4 Heisenberg and wheat grains

So we have a networked, very large-scale communication system in which every contact is stored and in which it is possible to read out the entire memory abrupt with every contact. The memory has not changed. Everything happens on a level that is not accessible to us, even below Heisenberg's uncertainty principle and in times of $10^{-18}$ s. The matter concentrations should initially be limited to the size of a solar mass, i.e., $10^{30}$ kg or about $10^{57}$ particles, which all belong together and are interlinked.

Statistically, we then have $2^{10^{57}}$ possibilities in principle for arrangements between the particles. In a chess game, if one put twice the amount of grains on each square, we would have 18 billion times billions of grains on the sixty-fourth square, far more than there is wheat on earth. Here, however, it is not $2^{64}$, but rather an incredibly high number. This means that the whole thing must have been reduced from the beginning, but the number of possibilities to link the particles with each other is infinitely large.

So, we have a highly interconnected, large-scale excitation system and an unlimited number of possibilities. How do we find an evaluation of the connections that makes the system more complex in a positive sense? Who sorts out the right ones, the valuable ones, from the flood of possible arrangements? How can conscious thinking and rational action arise in it? If the system is in a permanent state of excitement and can easily build up, i.e., if there are repeated resonances, which is certainly the case in our arrangement, then we have everything except the most difficult, the selection or evaluation process. What in this universe tells the system this is good and that is bad? Why does complexity seem to arise in our world?

We ourselves know very well about the value of intelligence, consciousness and generally higher life. We know why it is worth spending 20% of energy on brain power even though it is only 2% of body weight. Five-year-old children even have a need of about 43%. The difference between being alive and feeling alive, consciously experiencing the feeling of being alive, is so special that we are happy to spend only for our lust, one fifth of the energy. Apart from this, our superiority has changed dramatically due to our intelligence on earth. We can influence life on Earth, the climate and maybe later on cosmic threats like asteroids. We have the potential to destroy everything and have made all big animals and plants more or less subject to us. We use the thinking and know of its value. But how can a dull, highly interconnected system know about it? And if it knows about it, who always sorts out the right approaches?

# 5 The search for the magic

Even the largest networks have a beginning. In our model, we even assume that a single, random first particle, the trigger particle, is responsible for a flood of motions of other particles, which then start the network in a chain reaction around the first trigger particle. Also, the range will probably be very large, but because of the finite speed of light, it will not go as far as it wants into space. The trigger particle creates something like a centre and the speed of light and the distance to other trigger particles creates a boundary between the networks. So, our networks are large but not endless. There are particles that belong to it and there are particles that belong to other networks. Also, some particles are closer to the centre and others are further away. So, we have a delimited space of the networks, but still no intelligence or conscious exchange. After all, the flood of statistical possibilities will be severely restricted. There is a belonging and there are connections to other networks to which one does not belong. These connections, this common ground, will be strengthened over time and with the ever-increasing exchange. Moreover, we have now found a reason why matter contracts. The exchange of information leads to an attraction, the particles are no longer homogeneously connected with each other across the whole universe, but there are areas that communicate with each other more intensively because they belong to it and there are areas with which less communication takes place. The particles slowly start moving towards the centre, the density of matter increases. Apart from this, the density of excitation also increases because the distance to the particles becomes smaller. More and more areas come closer and closer and have contact; so far so good, but there is still no intelligence in them.

We consciously describe the exchange with communication or terms such as community, even if this should only be purely physical. If we seek life, we must not treat particles and networks like number systems: Nothing interesting and exciting ever results from abstract connections and sober values. There must be something different from the beginning, something unexpectedly special. After all, we are not looking for the logic behind everything, but we are looking for the magic. When does life come into it, when is the whole suddenly more than its parts?

Our brain is very versatile and sees in things what it wants to see. If we describe wheat grains technically according to their number, their weight, their value on the market or their physic-chemical properties, then wheat is also something technically sober for us. If we describe a grain of wheat as the sum of certain molecules, atoms or quantum mechanical states, then we again encounter exact mathematics and statistical possibilities. But if we get involved with life, the liveliness that is in it, then we have to choose different words and then the brain sees different pictures than in scientific description.

So, let's stay with the communicative exchange and with our networks, which are excited. Let's put ourselves in the elementary particles that belong to a network, store everything and have large, versatile exchanges with other particles. All contacts are similar and yet different again. At each contact, the particles see the stored contents of the other particles, what is equal, but no two contents are the same. Each particle is individual, has made its own experiences. Here we find a dramatically different approach to quantum mechanics. In quantum theory, particles can assume the most diverse states, but all particles are identical. The states are not fixed, but arbitrary. Every particle can take on or lose any state, there is no individuality.

Our particles, however, are all small individual personalities due to experience and stored knowledge. As in the beginning only very banal things can be stored, like the position or the number of contacts to certain other particles, but in principle there can also be much more, much more important things stored. For example, at some point the knowledge of life. The genes are large, the DNA in humans weighs about one picogram ($10^{-12}$

g) and thus has about 3 billion base pairs. One base pair manages to store two bits, which are four states and thus 6.54 billion bits or 800 MB. This number is large, but is it enough to store all the knowledge or does it only create the condition, the base? Let's maybe just go further. The genome consists of protons, which in our view also create networks. A DNA would then contain $6 \cdot 10^{11}$ protons, i.e. 625 billion protons. However, this network does not store 4 states per base pair, but in principle $2^n$, i.e. $2^{625,000,000,000,000}$ states, which is again a number that goes beyond anything imaginable. With this, the position of every single atom can be stored, indeed of every atom in the whole universe. So we have two types of storage, on the one hand the world of DNA with its base pairs, which is familiar to us, it is comparable to our ideas of storage building blocks. And then there is apparently a second, much more unreal world of elementary particles themselves in the familiar world, which probably also functions completely differently, according to strange laws. And there we ask ourselves whether life is not stored in it. In elementary particles the positions to other atoms are stored but they also belong to the DNA and are therefore part of the knowledge of life. Every single one belongs to it and in the sum we have then possibly the wholeness of life. Or don't we? Or is it still completely different in the sum? Is there something missing from everything if it is to work, for example the outside world, the connections of DNA to other networks?

The whole thing is slowly rising above our heads and we are losing our perspective. So, we go back again.

# 6 Mathematics and quantum mechanical particles

We know that this knowledge must be large, because as complex and multi-layered as we are, there must be a much larger knowledge and network behind it. We also know that our building blocks have to be simple, otherwise it cannot develop from within itself. So maybe the secret must be in the network somehow. But how can a network find out what it wants? How can a network ever be more than a process?

For us living beings, the matter is immediately clear. If we had the prospect of living like particles for the rest of a hypothetical infinite life, we would quickly come up with something entertaining. Always trying something out again and again, always pushing the limits of what is possible, always looking for something new and exciting, the more complicated, the bigger the better. We would suffer terribly from the monotony. And assuming we could live forever, it would be a horror to be surrounded by a boring world. So, our world is as it is:

colourful, diverse, exciting but also beautiful. We wage too many wars and destroy so much, but never does a war ruin everything, and in the time afterwards it often goes on strengthened. We provide the networks, if they exist, with the entertainment. We don't want to be bored, so why should the networks want to be bored? If they created us or helped to create us, and perhaps are still involved today, then our world certainly says a lot about the consciousness of the networks. Then we are the fulfilment of a dream.

That means we are thought by big networks and these networks have intelligence and consciousness today. But where does it come from with the networks? Could it be just pure boredom that made the networks become more intelligent? Can particles get bored? If a particle has a purely quantum-mechanical structure, then it can be calculated exactly according to analytical formulas, states can be assigned to it, its uncertainty can be determined and statistical predictions can be made, all within the framework of the equations very precisely. But the essential thing is that the particles are interchangeable, i.e., there are no special particles; one is as good as the other. Such particles will not be bored. Not today and not in a billion years, no matter what the environment is like. Quantum-mechanical particles are born out of mathematics and just like numbers, they are soulless and empty. To feel boredom, we have to feel something must be something in us, we have to evaluate the processes in time. First of all, we have to notice that something repeats itself all the time, so it needs a memory, a memory. Then it must also know that there are other processes, that something can also be more versatile and varied. For this it needs the exchange, but not only of individual information, like: here I am and there you are, but the exchange of knowledge.

Among other things, knowledge stands for experiences that others have made. There is always a chronology in it, i.e., sequences or longer sequences are stored. For a more complicated knowledge not only pictures are enough, we have to store a temporal processes of pictures. For every static image, an actual state, the time must be recorded. If one then lets the pictures run chronologically, something new emerges. Movement is added, and with it an information that arises from movement. If this succeeds with elementary particles, then such particles would also be bored, and then we would immediately have a mechanism that ensures that change happens. In the fight against boredom then a search for more and more interesting challenges would result and then it is quite conceivable and not at all surprising if intelligence and consciousness would emerge. So, in our case the elementary particles would not only have to be communicative and able to store everything, but the exchange of information would also have to be more comprehensive than expected. The exchange would have to be abrupt, because only $10^{-18}$ s of time is available and this information must do something with the particle, a reaction as a whole must result. The

## 38  6 Mathematics and quantum mechanical particles

stored information together with the own information must lead to a movement which is connected with it. If there is such a significant information in the elementary, then the networks that are created with such particles can be so much more. Then such a network would not even be comparable to a computer. Because the most elementary is not a zero or a one, but the building blocks themselves are already immensely complex and contain the seed of life. However, it is not images as we know them that are stored in our particles, but rather only positions in space, if at all. Instead, the particle does not only learn the position of a single particle during the exchange, but when it reads the whole memory, it learns something about the positions of all stored particles of the other particle. If it reads the whole memory, it gets a momentary picture of data that was stored one after the other.

It is still very difficult to imagine all this, so let's first take an analogy from our world. We also tell each other stories that we have experienced and try to use a symbolic language to create images in the other person's world, or rather temporal image sequences. For example, we talk about someone we have met and if the person we are talking to knows her, then pictures of the person in question are created in her, but also the corresponding emotions, which are also stored. If we talk about a conversation, the other person might remember similar conversations. The more familiar the conversation partners are with each other; the more resonances arise and the better the conversation goes. The other person may learn new information without really having met the person being talked about. Communicative people can get a lot of information very quickly without having to experience them all, but we all have to get to know many different people and store their characteristics together with the emotions they feel, in order to feel a diverse world within us. Then we can only hear from the most different other people without being with them. Then only abstract words or letters in a text are enough to make the corresponding images jump up in us. When we are surrounded by many images, for example through films, spend a lot of time watching films, meet an enormous number of people every day due to our mobility and urbanization, we are almost happy if the exchange is only done through short messages via smartphone. Then we don't have to store so much secondary information and we can fill the content of the short message with pictures ourselves. These small text messages then have something calming in their abstract simplicity, they create structure. So, if we have a great world of experience within us, we can also only talk with abstract words and signs, only read a book and find it exciting like a film. Yes, computers even manage to store and transfer information temporarily only in zeros and ones, i.e., in the smallest possible unit of information. Nevertheless, we see how well and effectively this works. Conversely, we don't understand the language of our counterpart, if it is for example a foreign language, then it doesn't trigger any images in us at all. A person who does not know higher mathematics, for him the Einstein equations are only hieroglyphics. Einstein's equations show how long one has to be trained to understand such symbols.

This means that in principle, the information stored one after the other in an atom can also contain images and more complex information or can be assembled as such. Only, for computers many algorithms and formulas were developed to be able to store pictures, movies and other information with ones and zeros, intelligent brains were necessary for this, we don't have such a thing with the atoms. The elementary particles are if only the memories: the pictures and the knowledge must originate somewhere else. So, we assume again that this happens in large conscious or intelligent networks, and that it happens from within, without outside help. How is this possible?

After all, the images and emotions are not stored in us like photos and movies or writing in books, but very volatile. It pulses within us and we are only given the feeling of having seen and heard everything. If we open our eyes, we see the stable static reality and hear suitable sounds. If we close our eyes, we think we still have reality inside us, but it is much more unreal. Only sporadic things can be recalled, only fragments, but the thinking is not stable like a written story. Our memory is organic and alive. Maybe some networks within us, like quantum computers, calculate enormous amounts of data in a very short time and spit out plausible explanations and solutions that match what we have seen and heard, thus creating a tangible world of experience of the outside world within us. It seems that we are surrounded by two worlds, a sluggish, very slow and static world, in which connections are laboriously built up and stabilized with large atomic chains and molecules, this creates a scaffolding from which we can start, a foundation for stability. But then there is a second world, that of the quantum, which must have somehow managed to find ordered structures, although in it everything happens almost simultaneously and almost without distance playing a role. A short blinking of units of knowledge that are suddenly there, completely isolated and without the inert matter completely helplessly lost in space. Only the combination of slow, stable and inert processes can turn these spaceless thought images in the background into something meaningful for reality. Just as we can't do much with a quantum computer alone, but we always need our familiar computers, so it could also happen in the universe. A complex inert networked world of matter superimposed by quantum-mechanical worlds of thought.

# 7 Computers, switches and transistors

The conjecture is very exciting, but if we do not know how the networks of particles work, how they think, we are still no closer to the solution after life. Perhaps we first take a closer look at how a computer is constructed. Maybe this will bring us closer to the solution of our problem of understanding a particle network.
In classical computers, data is stored and processed in a way that is familiar to us, the only difference being that today everything happens in an insanely small space. The basis of a computer are small switches. Like an electric circuit with a lamp and a switch, it means the switch is closed 1, the lamp is on and the switch is open, the lamp is off 0. The switch is now replaced by electronic components, the transistors. For computers, mostly MOSFET transistors are used. Transistors are made of high-purity silicon, which is either doped by specific impurities p, i.e., has a lack of electrons and thus has a slight positive charge surplus, or the silicon is contaminated with phosphorus atoms, for example, and then has an electron surplus, i.e., the silicon is n-doped

or has a negative charge. In a MOSFET transistor there are two n-doped areas which are separated by a p-doped gate. The current is to flow through the two n-doped areas later, one is called source and the other is called drain. The gate in between now controls the current between source and drain by applying a voltage. If no voltage is applied, the two areas are isolated and no current flows. If a small voltage is applied to the gate in the middle, then a current flows. So, we have a switch that is controlled by a small electrical control current. Switch off 0, no current flows and switch on 1, a current flows. Because we no longer have any mechanical parts, we can now do a lot of switching in a very short time. A MOSFET transistor can open and close the switch up to 100 billion times per second.

Furthermore, it has been possible to build such a transistor on a very small scale. In the i7 processor from Intel, transistors with a size of only 22 nm were built in. This makes it possible to build processors in which 1.4 billion transistors are housed on an area of only 177 mm$^2$. That is only 1.33 cm edge length. These small switches only know 1 or 0 states, i.e., 2 bits. This is why the language of computers is binary. We humans calculate in the ten-system, which is due to the number of our fingers and thumbs, computers work in the dual system, but both are equivalent, only that the number of digits increases faster. For the circuits there are now logical connections if one connect several such switches with each other. Mathematically, this is called statement logic. The simplest circuits or logic gates are 'and' and 'or' blocks, i.e., two switches connected in series or two switches connected in parallel. In the first case, switch 1 and switch 2 must be closed for the lamp to light up or for the current to flow. In the second case, the 1st or 2nd switch must only be closed. There are also NAND and NOR and some other gates, but the RS flip-flop gates are especially interesting for us. With these gates, state 1 can be held permanently without really having a memory.

Although we have talked about the fact that we can store an endless amount in our atoms, this particle memory is fixed, we cannot write and delete it again and again, each contact is stored once and can only be read out but not changed. So, this is a pure static memory where knowledge about other particles is stored chronologically. However, we cannot invent intelligent processes without storing something in between. And this storage must not be static. For fast thinking we therefore also need fast memories, which sometimes hold intermediate values or results for a short time and sometimes longer until they are needed, this is provided by the flip-flop circuit. A flip-flop circuit consists of two NOR or two NAND components, which are connected in such a way that in a feedback of the values the state 1 can be switched on or off again by pressing again. Like a push button switch, where by pressing once something is switched on permanently and by pressing again everything is cleared again. Capacitors or coils can also be used to add a time delay, but this could also be done with inertia or distance to other particles. With a flipflop circuit one can set the 1 and one can delete it again. Both are electrically determined and controlled and in shortest time and that is important. If we only heard about these simple few logic elements and how many of them, we can fit into the smallest space we would never believe that we could ever do anything useful with them. Only with small electrical switches in endless numbers and a simple statement logic placed over them. But as incredible as it is, these computers work just as well and not only as machines as tall as houses, but also as wafer-thin tablets or small smartphones. In connection with the Internet, the younger generations can no longer imagine a world without computers. Just as the older generations could never have imagined such a digital world, when they were children. Perhaps there are parallels to our networks of particles. After all, the world of bits and bytes is also highly networked. We have large numbers of switching transistors involved, we have designed them in such a way that we can control them sensibly, so that they can handle tasks that also affect our macroscopically living world. We can carry out complex operations with the simplest means in the smallest detail. Using software, we can control the elements in such a way that they are easy to operate and still run complex programs on them. Here we have the architecture of the hardware and there the familiar machine language of the software and the interaction of the two are working. Without some people knowing exactly what the others think up and assemble, without even knowing why it runs so smoothly, we see the amazing result. In fact, millions of colour points can be controlled and reproduce a meaningful picture or film with sound in the highest quality. It may be that the first rough matrices were still so clear that one could understand what was happening. In the meantime, we claim that, according to known mechanisms, the whole thing has been reduced more and more and the density multiplied without knowing exactly what and how it happens, but it works. The process of miniaturisation works so well that billions of transistors can now be accommodated in a sensible way in the smallest space. If one connects the processor, the programs run on it, only better and faster, which is almost miraculous for an outsider. It is as if matter is just waiting to digitize the world in processors to be able to process and record images and stored knowledge in a stable way. Something that is created only fleetingly in our fantasies. To store something exactly in pictures, films and knowledge and to bring it out again one hundred percent at any time without loss. To stop time, to let it run back and forth, to pull pages, pictures, information with the fingers, big or small. And all this not only as a roughly like this, but in brilliant colours of unrivalled resolution and best sound quality. We all know what it means to store knowledge. Gutenberg has shown us how to disseminate knowledge en masse, static knowledge.

Film, television and radio can store everything we experience and play it back in completely different places at completely different times, but with computers and the Internet, stored content is no longer a one-way street. Now it is possible to actively influence everything in the present time, to have a connection to any place in the world, to change pictures, to let fantastic worlds only arise virtually. To play games online with others, where the end is open, but where the world seems alive and dynamic, even though it is not real. In these worlds, there doesn't have to be gravity, there doesn't have to be stability, only what is thought can seemingly become real in them, free from the heavy physical laws. Life is a game with the lightness of abstractions. Maybe this is what matter wanted and maybe much more and that is why it has been able to develop so successfully at this insane speed, as soon as it knew how to do it?

# 8 The rhythm of the atoms

We know that in networks there is communication, exchange of information, memories that can be written and erased, a system of evaluation and classification, structures that function according to logical laws. All this requires energy, a continuous temporal sequence between the objects in space. Quantum-mechanical objects are in constant motion, but do not constantly exchange information. When they exchange, it is always a measurement. In doing so, they change the environment and are themselves changed, they become concrete for the moment. The sum of the states is determined at the time of the measurement. In addition, they move in gravitational space on bended tracks because space is bended by masses, not because they are in connection with other mass particles and thus feel a small attraction. A particle of quantum mechanics can fly past a large mass without the many atoms of the mass accumulation knowing about it. The motion of the individual particle can be passively curved, i.e., it can be influenced without anything being exchanged, without a measurement being made. Quantum mechanical particles can sneak through the universe secretly.

For our particles, this would be impossible. Not everyone knows where a single particle is, but each particle has constant contact with various other particles, nothing gets lost or can sneak away. We also feel that such a continuous stream of information is part of the liveliness. Particles that are only here and there for a short time and unpredictably, how can we build reliable networks with such particles. In our case, the basis is an intrinsic cyclic exchange between the two charges, which are always as similar as possible, to the most varied masses and to the antiparticle, which lies on the edge and moves away at the speed of light.

Quantum mechanically, the world takes place predominantly actively in the electrical domain. When we speak of a measurement, we always mean an electrical contact. Gravity influences the particle only passively and inertia is also produced by a field, a scalar Higgs field, but this is not a measurement either, something that has to do with our world. In our picture of the particle, they are integrated into a sequence of events, our particles tick incessantly, have their contacts in a continuous sequence. No matter whether they are still dull or already alive, we have something that ticks like a clock or like a heart and that sets the basic rhythm. For particles that are too mathematically statistical, life must come from outside, from above. Something must create it, someone must connect the building blocks in a meaningful way, supply them with energy and then bring them to life. Dead little spheres do not do this by themselves. Not today and not in billions of years. That doesn't mean that there can't be such an exterior that brings matter to life, but then we'll be with higher beings and gods faster than we'd like to as scientifically thinking people.

If, on the other hand, the atoms themselves pulsate, if there is already a cycle between the particles from within themselves, even in the case of the electron and the proton, then we also have the chance to build up life from below. Because we intuitively feel that, like a foetus that can build itself up into complex life through permanent division, the parts must all be connected to each other in a meaningful way, and in complex cell structures there must be a rhythm, a ticking heartbeat, something centrally connecting. All this must also be present in the atoms if they are to join together to think.

Nevertheless, we are a small step further, because we realize that for life there must be this continuous connecting current that our particles bring with them. From this, in analogy to the foetus, a second ticking process time must then unfold, which establishes the basic rhythm and togetherness for a limited number of atoms.

The hardware in a computer is the prerequisite, but without a program, without the software, nothing happens with the computer out of itself. In this case it seems that first the hardware was there and then the software. First one had to see what could be technically realized and then one developed a suitable programming language. In our world we often have the feeling that there were first the accidental inventions, like the wheel or the book press and then a revolution in thinking arose from it. And of course, we always need the real objects to be able to try out our imagined thoughts. But it remains the case, if we want to replace a mechanical switch, a mechanical relay with an electronic component, our mind must be ready for it beforehand. If one put an ignorant person in front of a circuit with a transistor, it does not understand anything. It is only by observing round materials that motion sequences can play out in the brain, which leads to the discovery of the wheel. Our brain or something in us is apparently predestined to recognize processes and to be able to play through ideas in our head.

So, the hardware became more and more perfect but also the software became independent and bigger and more and more complicated complex programs were developed and made to run on the computer. But it has not yet been possible to discover life with them. Even though we are increasingly successful in developing cyborgs, i.e., living creatures technically improved, we still need something living from nature to live. Although complex programs develop a certain momentum of their own and we can program an amazing amount of intelligent behaviour from the outside, but it is always artificial, it is always our thoughts that are reflected in the machine, even if we no longer recognize it after a certain complexity. So, all this happens from the outside, because it was designed that way from the beginning. How can we let it develop from the inside, from below, even if it is only in our imagination? How can a single communicating particle, which also keeps everything it has experienced, recognize itself in any way? How can it understand, evaluate or even influence images or knowledge of other particles? With a transistor or a switch, the matter would be clear, it doesn't work, not by itself, but does it follow that particles can't do that either? Not even a hint of understanding?

Another question is whether it will ever be possible to program life into the computer, that is, to bring life from above into the computer, and the answer is not clear. If consciousness developed out of itself with simple particles, then it must also be possible in principle with sufficiently highly networked transistors, just created from outside. Thinking cannot be that exclusive, that only another highly intelligent being can create something like that, can it?

There is the big difference between a transistor and a particle, that transistors are more like quantum-mechanical particles, all uniform, so they don't know boredom. After that, no matter how big networks of processors we could switch and power them, they wouldn't get smart on their own. If so, then the networks would always have to be sensibly linked, the liveliness would have to be programmed in. But what is missing are the long evolutionary developmental steps over time. We would always program something that has something to do with us, abbreviate evolution, but also not create something amazingly new. If we did, we would get a copy of ourselves, but maybe that is what we just want. But that leaves the doubt whether such an artificial intelligence is ever independent. Whether it can ever be more than a highly complicated machine. In the case of organisms, we have the feeling Organic beings are diverse and always different, each one seems to be independent. Organic life seems to be the more highly developed it is, the more complex its structure is. The higher life is not only on one level of life, but at least two, if not three different levels of complexity. We have the ego that arises from the body cells, then there are the microbes, without which we also cannot function, and then we have the particle level, which cannot only be used passively like building blocks, but has its own life of its own. We also claim that in addition to the particles that belong to the body, there are also extremely large external particle networks that can also think and have consciousness and that are in communicative exchange with our body particles. These large networks do not only control us humans, but they are always present with all living organisms in the world, plants and animals, but also with all domains of life, with bacteria, archaeae and eukaryotes. The only question is: are we then only copies of this thinking or are we even only controlled like puppets? Are we also no more than what an artificial intelligence is for us? And if these intelligent networks exist, where are they? Aren't the distances in the universe much too great?

What do we get out of it if consciousness is located in the centre of the Milky Way, then every communicative exchange would take 27,000 years, because the centre is so many light years away from us. We are going in circles and want to solve too much at the same time again. Let us stay with our particles. Elementary particles are not quantum-mechanical particles in our image, because they exchange continuously, store everything and can read the memory from foreign particles. There is a course of time, a rhythm of time and there follows a movement of the particles, which is connected with the exchange to other particles, but we do not believe that the particles themselves are already alive. No particle is like another, there is a heartbeat and communication, but where should the information be processed, what should it evaluate? Particles bring everything for life with, but in our opinion, they do not have their own independent life in such isolation.

From our macroscopic view of the world, particles still behave quantum mechanically. They can assume different states, we cannot distinguish them, their whereabouts cannot be clearly determined and two particles connected to each other are apparently separate and yet together. In the quantum-mechanical world the atoms are running for us, it is the world we can observe. But it is not the deepest connection. The space between the particles does not really exist. It's abstract. At each contact, elementary particles are connected to the exchange particle for $10^{-18}$ s, no matter where the other particle is and how far away. One contact is spatially close and widely distributed via gravity and a second contact in the cycle concerns the charge, which mostly runs off in close proximity to the same counterpart. We call far contacts which then also occur over a somewhat longer period of time quanta, for example photons in light. The particles are also connected to each other without space, no matter how far away they are and that not only for $10^{-18}$ s but for the time length of the photon exchange. Consequently, a superimposed movement still takes place. The information can then be longer and more complex. These quantum connections are the ones that physicists have explored and on which the world views are built. In contrast, the gravitational connections are too weak and too disordered and can only be described as a whole in the theory of relativity. It makes an infinitely fine space and an infinitely fine time out of the vast number of individual connections, which are curved by large masses. It is the summary of the effects, but it is not the mechanisms themselves that are covered here. And just as we cannot observe the details in gravity, the same cannot be done of quantum mechanical states and connections. It is only because the electrical connections to other particles last for such long periods of time that we can detect patterns that result from these contacts. We do not have a view of the networks, but of what results from them. And, is it only life that is in the electrical connections, do we have to continue with quantum mechanical particles or is the secret in gravity? Do the masses form the great networks that have discovered thought? Is it the electrical, i.e., the resulting quantum-mechanical stable connections that order and bundle the many, or is it the extremely simple gravitational contacts with their simple information, which can however become gigantic. Do networks only have to be large enough, then special structures automatically result or things must be stabilized and combined to an order scheme. Or is it, as so often, a mixture of both?

If thinking would also be possible only by gravity, why do charges exist? Can it be at all possible that any property of matter is superfluous, not needed at all? It is conceivable that matter alone creates thinking, but that complex life is then not possible. We would have a thinking conscious universe, but also a very boring one. After all, the electrical connections make the kind of complexity possible, which in itself creates diverse, colourful worlds that are exciting and unpredictable. So, our earth certainly needs both to be able to come into being. And if there is this strange second thinking in the universe, then we are at least as important for this consciousness as the thinking that created us. Then this consciousness is not only in some way connected with our consciousness, but then all our cells and particles have contact with it.

In any case, the consciousness of the networks would then be the much older one. The universe is about 13.8 billion years old. Our earth, the mass of the Milky Way and the surrounding area must belong to the oldest matter and, according to our ideas, was formed a few million years after the beginning. The earth and the solar system are about 4.5 billion years old, which is much younger than the age of the universe. Since we believe that already thinking networks were involved in the fusion of the elements and the preparation to form an earth, consciousness must therefore be ancient. Perhaps the first controlled processes or other pre-forms of intelligence emerged relatively soon, as matter contracted into stars. And a further assumption is that the large mass accumulations in the centres of galaxies, the supposed supermassive black holes, are much more important and more than just matter destruction machines. That the matter in this area does not disappear, but occupies the space so compactly that already again an order stabilizes the particles and thereby not only the free networks calm down, but the matter crystallizes to certain patterns matching the networks themselves. The little free space forces the neutrons to be stable almost like a solid, only that now even the possible quantum mechanical states are all used and every particle has its own state. Each particle in this endlessly large ordered collection could then represent a qubit that is not continuously connected to our world via collisions and photons. Instead of an all-destroying black void, there could also be the centres of consciousness of our universe. A huge quantum computer, in search of complexity and diversity.

# 9 The entropy in the networks

But we digress. We are still searching for the most elementary level of intentional action. Electrons and protons can be networked and store information in an elementary way, but they are unlikely to evaluate their own content. The difficulty to get further here lies in the simplicity of the objects. How is it possible to recognize the germ of intelligence and consciousness if electrons or protons are considered specifically, then they do not seem to be more than just numerical values. Nevertheless, it is their properties that create networks and connections. And we already suspect that something special will be added at the latest in the networks. If we therefore make a different attempt, we can assume that the networked particle region is contracting.
In a first step, huge areas of space, triggered by a single trigger particle, are separated from each other. The areas can have dimensions that the mass is present for whole suns within. With the size the delay by the finite speed of light already plays an important role. This means that with increasing compression, more and more

distant particles come into contact with each other, but it is also easy to imagine that the one large area of space fragments further. Smaller and larger islands are formed, which are more or less connected with each other. The excitation density increases overall, but especially in the small subareas. The islands create family units. In them, the processes run very fast and become individual very quickly, if only because every particle in them is also individual. But there is still contact with other islands. Perhaps in such large volume areas further subunits are formed, different levels of connectivity. Up to now we have apparently living units, from the outside there seems to be a lot of movement, but there is no directional approach, no selection process that prefers any kind of path. But we urgently need one, otherwise the islands will exchange ideas without anything special ever happening.

Mathematically and statistically, the particles are only told that they increase the number of possibilities from within themselves, never decrease it. The degree of order can only decrease afterwards. Physically, a more disorderly state has a lower level than an ordered one and all closed areas always strive for the lowest energy state. This applies to particles that are small spheres or particles that are quantum mechanical objects. Both are mathematical objects. Our particles, however, network from within themselves, which does not necessarily lead to a contradiction to entropy. The number of possibilities increases due to networking. But if an order is formed within the network, this would be a violation. The networks must remain mathematically confused. But that's where the problems start, because nobody can follow these networks. They are huge, extremely light and lie below quantum mechanical measurability. This means that we do not know whether the second thermodynamic law of thermodynamics also applies to the networks between the particles. We can only measure the coarse structure of mean values of such particles, which results from this. And this only concerns the electrical connections. Gravity, which is much more widely scattered and only shows itself in large mass movements, cannot be measured individually for a particle at all.

If we also want to explain life, liveliness, then we have to give up entropy for the networks below quantum mechanics. The cool abstraction of mathematics does not seem to rule here. Here, where energy only plays a subordinate role because it is in abundance, the networks seem to follow different rules. Here, perhaps the fight against boredom applies. The networks of particles do not seek the simplest and energetically most favourable way, but on the contrary, they seek variety. They are looking for the complicated, the complex, the higher the better, and for where something is going on, how the matter has to be arranged so that most things happen. The fight against monotony. Alone, particles are nothing, but in groups everything is always in motion, always multi-layered and complex. From our point of view, these movements are very fast. Many processes happen in just a single second. If this were to happen over billions of years, it would be dramatically monotonous. If we now pretend that networks are always looking for the complicated, the more variety, the better, then suddenly everything is open again. Just as disorder automatically increases in our world, but also the number of possibilities, so in these networks the search for the greatest possible complexity could set the tone. Then, when energy plays no role, completely different structures are formed. Then also the liveliness from the complexity has become tangible. The energy or the size of the mass is in the distance between the levels. If the planes are closer together, the mass is larger or the particles have more energy than if they are further apart. This takes place within the distance Re. The planes can shift in delta steps of $10^{-57}$ m, which are tiny shifts or tiny amounts of energy. Each contact shifts the planes by this small amount, so the particles lose a little bit of their own energy and absorb the same amount of external energy. But the energy is now fragmented and distributed throughout the universe. Such a thing could be compared to entropy, particles bring compact energy with them and distribute it over space, the number of states increases. Only, now they also get energy from the other particle, so it doesn't flow only out but networks emerge and from a boring state a much more interesting one is created. If we all save our money, it is nicely ordered, but if we use it, machines and factories can be built, then it is productive. If, in addition, nothing can be lost in the process, but only transformed, then we would have gained a great deal by using it. In the particle realm we do not lose energy to space. If we have a fire in the oven, we get radiant heat, which is valuable, but when the fire has burned down, the energy is gone and lost. First the room was heated up, then the heat flowed out of the house and warmed up the surroundings and then space became a little bit warmer, only that it doesn't care and it can take up heat endlessly. Here the entropy only increases and the energy from space never comes back by itself. With elementary particles the energy is also distributed to more and more particles, but only in exchange and we do not lose something, but gain. Particles that were newly created and initially rested can now move. The range of motion becomes larger and larger and with each contact also more versatile. Such particles do not have then for a long time always the same speed and direction, before they finally meet another particle and change their direction. Instead, due to the increasing networking, they also get a change of movement and direction during the communicative exchange, i.e., no bumping, but only a connection that arises from the network and must be redeemed. This really creates movement; it pulsates without really losing anything. We see that particles can exchange billions of years without their inner fire going out. The older a particle, the more inert it is. The inertia has to do with the number of connections to other particles or the process time, how many processes, i.e., connections a particle can have to other particles per second and this is very high in our case. Therefore, the uncertainty is very small and in the case of electrons it is exactly at the lowest level in the atomic structure.

So, we do not lose anything here, we only gain. Already the movements of the particles show an incalculable versatility and no endlessly same movement in space. Entropy is not comparable with this. And also, the mathematical consideration falls. The number of states increases by itself, but in principle it is also reversible and nothing is lost. Moreover, from our point of view, from our evaluation patterns, versatility is much more valuable than uniformity. Furthermore, networked moving systems can also build up, there are resonance points in them. This cannot be done with single particles. This means that a networked system with so many particles have an extremely large number of possibilities to build up. This will certainly bring a lot of momentum to the networks. Assuming that resonances are played out in these islands of networks until a more complex, stable connection develops, then there is the possibility to tell the other islands about it. Others can then also achieve this more complex form. If this form later leads to stagnation, then it disintegrates again and new paths are sought. If we just assume that such experiments would be remembered and the most different ways would be compared with each other, then we would have small researchers who are searching for the really big complexity. The elements, the prerequisites for this are there, but how it should actually be done is still in the dark. Such small researchers would make everything quite simple because they come from our familiar world.

In the world of particles, we have the networks that are so different from our world, that can try out so many things in a relaxed way without being used up. This networking of matter is extremely large, they are very fast, there is a lot of exchange and there are ways to store found things. They are probably still mechanistic, so soulless, but do they have the germ of life in their pattern? We too can still be mechanistic, but it would be impossible to catch on to this high level of complexity, we are so interlaced, networked on so many levels, who would want to discover the machine in us. We are animated, but not only us, but also the plants, the animals and much more.

And yet we can't pin this particularity on anything in particular. We only notice that everything is always connected to everything else and that this does not stop at the micro level. Even the atoms are much more than just mathematical structures that function according to physical laws. And isn't that exactly the essence of the soul? Something that comes on top of the worldly things that can be explained? Neither do we great beings' function like mathematical numbers, nor like an extremely complicated calculation, we all know that, but neither do our atoms. Just not because they are so quantum-mechanically free and yet uniform, but because they are connected to so many other atoms of the outside world by inertia. The atoms are the gateway to the really large networks and, as we suspect, to other thinking. Their vagueness is a measure of inertia and the movements within it is an expression of the connections. A proton in us is in contact with a far away proton and exchanges memory. It then changes its speed and direction according to the information. There is the external motion of the proton which is related to our body with our momentary velocity and superimposed there is the proper motion of the proton, which seems to lie in a statistically normal distribution within the blur. Only this motion, which makes the mass inert, is part of the outer network and is full of information. Tiny packets of information are stored for a tiny little moment. Much shorter than the comparatively huge information packets of photons and quanta. We can analyse photons, they are big, but not the information of gravity, of inertia. They are too light and too short, but there are a huge number of them. The stream of gravity surpasses everything imaginable and cannot be compared at all with the stream of radiation from the sun, it is so much bigger. Only the electrical connections between the charges are equally large. The charges incessantly hold the components together in us and inertia creates the connection. Both together make up life. The difference between a stone and us lies in the quality of the connections. In stone there are only atoms of the same kind in a grid-like arrangement. In our body we find the most different atoms and molecules in a structured arrangement. Everything in us is knowing in some way. It knows what it must do to make the whole thing work. Every proton, every electron does not find itself by chance at the place where it is or does everything right for us by chance. That's why the inertia of particles cannot simply be free and random. The connection via the inertia of the animate beings must be connected with the other intelligence or the other intelligences. Something that knows where everything belongs. Whether it can also determine our life should remain open, but it might know what to do with the atoms in us and that is difficult and important enough.

# 10 | A diversified world

So, let's give it another shot. A particle has three different contacts. One to the antiparticle at the edge, one electrical contact to always the same charge and one gravitational contact to all other particles it had contact with. These gravitational contacts are stored, the electrical ones and those towards the edge probably not. The decisive thing now is the continuous process. It comes from the particle itself, there is something ticking. The particle has a heartbeat which is not consumed and which is always the same. It is intrinsic, like a spin, from the moment it entered the world and the antiparticle remote itself at the edge. It sets everything in motion and there is a stepped kind of spin. This kind of heartbeat causes the particles to move and intertwine, and the nets are regularly called up. This exchange still keeps a diverse world in motion today and makes this so simply constructed particle, with its two levels, something so special. We are almost inclined to call this passage of time, this ticking, the spark of life that makes the contacts come alive, only that it cannot evaluate anything. This evaluation of the nets and movements, of what comes out of them turns out to be a hard nut to crack. And if there is no evaluation of what has been created in our sense, there must be some kind of passage or filter that only allows certain connections. To promote the one and to sort out the other. What does that decide?

Assuming there is such a thing as the search for complexity in our world, how can a particle decide this path leads us further and that one does not? Surely it would have to be able to extrapolate processes for that. To look at the past, play through the future and from this develops the decision for the now. The charges create a stable framework, the solid structure, gravity takes over the function for the networks with their storage in the particle and the edge defines the limits of how far atoms can move away or how densely packed matter can become maximum. Matter bodies are slow and inert, quanta and networks are fast, almost timeless and light. If heavy or inert large matter is to be changed, then it goes endlessly slow. An agile mind could have gone through all possible variations many times. Over and over again. But only thoughts alone cannot move masses and masses that move undirected are too slow for thinking. There must be a connection - an interface for both. Matter must be modelled by thoughts. If there would be a connection then matter could stand for the reality of the world of thoughts, then the many possible ideas would be sorted out by what is possible. The physical creates hard limits. Conversely, if we did not have the time to try out all the possibilities with this slow, inert and heavy matter, then matter would still be busy today just collecting atoms and would be eternally distant from fusion. It would even be more likely that our universe is still only filled with a uniformly distributed gas.

Well, fortunately our universe is diverse and particles themselves have almost something like a rudimentary liveliness. On the one hand, distances are endless and inaccessible; on the other hand, they are time- and spaceless for the tiniest of time pulses. Like entangled particles, particles have a connection to other particles, which have more interesting and longer memories with each new contact. If we are allowed, we would say that there are contacts between particles, where they both feel comfortable because it harmonizes or something like resonance is created and there are exchange partners that cannot get along at all. Then we have described something soberly with our words and pictures and interpreted something into it, which might not be true. Are we allowed to transfer this; can it be something similar in the most elementary form as it is with us? Either we do it more soberly, more mechanically than we would like, or a particle actually feels something like preferences and aversions. If the immense flood of images it sees in the other person harmonizes, then the partners move closer, if not then they move away. One can describe it rationally and emotionally, but the number of stored contents is extremely large and the diversity correspondingly. Two foreign particles which then have contact and react to it with a movement have to do in both cases with a manifold number of possibilities. What one interprets then into it, whether particles feel something or only react passively is not so important anymore. The variety of possibilities to react creates many interesting networks and therefore a diversified world. And that is decisive. We could, if we wanted to, rediscover self-similarity in the smallest detail. All particles exchange, but all particles are also different, individual. Over time, worlds and contacts between particles crystallize. Which ones like each other and which ones do not; the smallest in the mirror of the largest. Only, then we would have something comparable with our life, but it seeks harmony, not complexity. A harmonious peace-loving world in the smallest would not develop, what for. So, the harmony has to be disturbed and there our boredom comes into play again. Just as in the Bible Adam was seduced by Eve with the apple of knowledge and both were expelled from paradise, our atoms also find an interplay in harmony for all eternity to be rather monotonous. Maybe it was long enough for particles to be close enough to each other, but something, if only time, made it clear to the particles that everything always happened in the same way. This is the moment of realization that knowledge about oneself, the monotony and the search for something new, the curiosity and dissatisfaction was born. Since then, the harmonious states of equilibrium are only intermediate stages to ever more and ever higher.

No longer the ideal happy world is the ultimate goal, but rather variety, the challenge, the difficult to achieve. The particles had to want this, they had to change the thinking in the networks accordingly, only then could evolution begin.

To this day, everywhere in nature the struggle for survival is hard and merciless. Even in humans, nothing has changed. But we do not only have to fight, we can also enjoy our happiness to be born in a colourful world. We can feel pleasure, release great emotions and make our lives varied and exciting. Just like children do every day. But we should not be surprised that a god allows suffering. For the individual, suffering is cruel and whom it affects when is random, no one knows beforehand. In this lies a great injustice in the world, but it would be worse if evolution had stopped at the amoeba, just because higher beings can suffer so terribly. We must age, at some point become sickly and shrivelled and say goodbye to the world again, anything else would be boring and lead to much too long phases of stagnation.

It is probably not absolutely necessary that joints or the skin and organs wear out with age. They could be renewed again and again in completely different periods of time and stay young. As a child, we cannot grow and become an adult fast enough and as we grow older, we all have to accept the decline of slow decay. Even if we are very attached to life, the end reaches each of us and it is seldom easy to end life, to leave the world again. But without this, life would not seem to be varied enough; we would be complacent and content too soon. If we would stay young and at the same time become very, very old, why would we still need children and why would we rush or want to achieve great things? Our patterns of thought would remain and no one new would make them collapse. Everything would flow the same way forever and ever. A peaceful, friendly monotony for all eternity and that, we suspect, is not what the networks, evolution or creation want at all. Happiness and joy yes, but not stagnation.

Beauty seems to be very important, the beauty of the world, its many colours, but also that of the individual, the variety and the many small nuances. One is never like the other. We do not have copies of ourselves; there is no other me, not even in twins. The image of the world of beings on it changes with time. No day is like the other, nothing remains over the millions of years. Even lust, physical lust seems to be enormously important; something that does not work in abstract networks without matter. A timeless fleeting thought without the weight of the inert mass, its fixed boundaries, which bring slowness and calm, is a lust, a feeling of pleasure not possible. But the feelings, the many small and big feelings in life are needed by us and all animals and plants. We even suspect that even with atoms and molecules, there are signs of happiness and unhappiness, joy and suffering. A particle that meets another particle again and again, where the stored images harmonize, is happy to be together with it.

# 11 The first consciousness

In fact, we suspect that the really big networks that think everything, control and influence us and have first thought and then created the complex life of something as central as the supposed black holes in the centres of the Milky Way. Perhaps the center, the consciousness for our life could be located here. Something that thinks, is intelligent and has consciousness, but is not beautiful and also not like us has feelings and is alive in our own way. It can think, but it can only build something somewhere else, create something and design something. It is the older, the more archaic first consciousness. We come much later and have a much slower, but mobile and more complex consciousness. Probably we are the second consciousness on which the first still hangs. It may well be that our consciousness functions independently, but is not free from the first consciousness. That would then explain a lot. For example, why we can ruin ourselves, consciously take substances that harm us, do things that are wrong although there is a bigger, deeper, older consciousness in us that knows what is right. But perhaps these foreign networks also love the irrationality, the irrationality, the open and new that arises from it. Only, if we were puppets of the first intelligence, then we would have to be controlled by infinitely knowing rational networks. Everything we have already said about old age and dying would then also be the case here. In no time life would be predictable and boring again on earth. So, it is much more favourable to leave us the freedom of choice, even if it is dangerous or deadly.

In spite of everything we have already found out how many possibilities are already in the particle itself and what enormous networks can be built with it, we do not suspect that life could be created here alone and independently on earth. We, the plants and animals; bacteria archaeae and eukaryotes are too complex, too multimolecularly connected. Such a thing cannot have come into being or function on a daily basis without help. It needs an evolutionary accelerator, namely thinking, something that can play through so quickly and so many possibilities purely mentally and does not have to try them all. Only then can something built up be created in a reasonable time, something that is stable and remains dynamic.

The biggest difference between man and animal is not the size of his brain. There are animals like elephants and whales or dolphins, they have much bigger brains. And even though we have the smallest nerve cells in the grey mass and therefore the highest nerve cell density, there are also some animals that are in a similarly high range as we are. But the real intelligence accelerator was the discovery of language. Language gave us an advantage in intelligence that is second to none. Now knowledge and experience could be exchanged, and in ever greater detail. If we try to think without language, we immediately fall back to the level of an infant. The second intelligence accelerator of man was writing and later the mass distribution of it. Today it is digitalization that enables us to penetrate ever deeper into the secrets of the world and to build the most complicated machines of any size. The computers and their networks make us as mankind so extraordinarily intelligent and superior. We often have the feeling that we can do everything, achieve everything. And according to our ideas, evolution also needs such an intelligence accelerator in order to first create life and later complex life from atoms and simple molecules. These accelerators are our networks of matter, possibly located in the centres of galaxies.

# 12 Entanglement

Before we delve into the centres of galaxies, we should perhaps go back to one of the most dazzling and strange results of quantum theory, entanglement. In German it is called "Verschränkung" and means something like enmeshing, intertwining or involvement which says more than this mysterious word "Verschränkung" or entanglement.
In quantum mechanics, many of the properties of elementary objects are summarized in states. A significant quantum mechanical state would be the spin of a particle, but also the spin of quanta, for example light quanta, the photons. There are various interpretations of the term state in quantum mechanics, but the Copenhagen interpretation has prevailed. It is based on the assumption that it is not theories that are imperfect, but that nature itself is not deterministic in the most elementary sense, and that therefore the processes can no longer be predicted exactly from a certain accuracy onwards, but can only be determined with probabilities. For wave functions there are only state functions for which the probability can be calculated, but which have only limited reality. States which are unambiguously

defined as measured variables are called the eigenvalues such as the location, the momentum or bound states in the atom, the energy eigenstates. These measured variables are then the average or expected value as a statistical summary of the possible probabilities. How ambiguous quantum mechanics becomes as a result is shown by the fact that two particles can be bound to each other and the states can overlap. Since we cannot find out the nature of the state without measuring it, we do not know what state each of these two particles has. For example, two particles connected in their states can be superimposed in their spin state. Quantum mechanically, two particles cannot both have the same state, so one must have the spin up and the other the spin down. If, for example, two energetic gamma quanta are fired at each other, a pair of particles, an electron and a positron, are created which are entangled with each other. They have opposite charges, but also the spin up and the spin down state. They are quantum mechanically entangled with each other in their state. There is nothing amazing about that, one will have the spin up and the other the spin down. But that's where the difference starts. Because it is in principle impossible to find out which particle is directed up and which down, both can have both. And now the physicists go one step further and say that according to the equations, each of the two particles has both states at the same time and maintains them until a measurement is made. Only the measurement makes the state clear. Before that, both remain possible until the end. And this interpretation is neither compatible with classical physics nor with our familiar ideas about the world. For us, the particle is not measurable, but it must have one of the two possible states from the beginning. It is only made visible by the measurement and not decided by the measurement. Let us continue to look at our positron and electron from above. The two particles move away from each other but are still entangled. If we catch one of the two, let's say the electron, and measure it, then according to the theory the spin state is determined at that moment. By the measurement the electron knows it has spin up. At the same moment the positron knows and we know that it must have spin down. But this was also determined directly with the measurement of the electron, no matter how far away the positron is from the electron. Suddenly and without any loss of time the positron is told that the electron had the spin up. An exchange of information without time and space. And only by measuring the electron does the positron become an independent particle. Before that, they were mysteriously not really separated from each other, at least not yet their states.

Here, our familiar world of distances and time delays is turned upside down, yet it fits in exactly with what we imagine in our model of the particle world at each exchange. Not only in the long quantum-mechanical contacts, as seen in the entanglement, is the exchange momentarily space- and timeless, but also in every extremely brief contact that particles have with each other continuously. Their position thus moves within the expected value, but it is only the statistical summary, i.e., the average value of these many individual contacts. In entanglement, physicists allow something that is astonishing and yet works, although it shouldn't surprise one that much. They also admit that high-speed particles close to the speed of light perceive distances quite differently. But in the case of electrical and gravitational exchange, which happens exactly at the speed of light and is therefore space- and timeless, they shy away and prefer to introduce classical fields that bring stability and calm back. All because one cannot trace the individual contact. Not that of gravity, but also not the individual between the charges, which is equal in amount.

In the case of entanglement, we note with amazement that the connection between particles can also happen quite differently. It shows us that we have a wrong picture of space and the distances in it, but the states between the two entangled particles are large and strong information connections. They are so strong that we can measure them with our macroscopic rough machines. But what can only be observed here as the tip of the iceberg also applies to the many information contacts below the surface, each of which is too light to be measured.

What has been explained here specifically in this example of pair formation actually applies to all quantum mechanical particles. As long as they are not measured, their states are still open according to theory. With entanglement this interpretation of nature is only obvious, because it is an isolated state that two particles or photons can take and this strange behaviour can be measured and proven very well. But one assumes that also electrons in the bound state are ambiguous in their quantum states until a measurement or a reaction with another electron or photon occurs. Only then the electron has to decide which quantum states it has. So afterwards in the normal state of matter we have a free, indeterministic world. At the smallest scale the objects are without a causal relationship; no wonder Albert Einstein did not agree with Nils Bohr on this. Nils Bohr describes the world from our point of view, and in it the objects do not seem to be causal as long as they are not in contact with the world, which according to his ideas is mostly the case. A contact can be a collision, but also a quantum exchange. But Albert Einstein also assumed from his infinitely fine curved space and that the particles in it have a reality that is related to our world. But if we assume that, in addition to the mean values of the particles of time and place, there are also connections between the particles that are extremely large in terms of the number of exchanges and simply cannot be measured because they are much too light, then a second world results. We know exactly that the gravitation is $10^{38}$ times smaller compared to the electrical connection. Even the forces that occur in a single elementary charge are very weak. If we wanted to measure the force of a single graviton it would be $10^{38}$ times smaller. And this is exactly wrong, because the electrical exchange forces are equal, otherwise the inertia could not balance the charges. The difference is that charges are essentially always directed towards the same pair of charges. If the measurement takes 10 s, then during this time the electron and the proton have exchanged $10^{19}$ times, had contact with each other. There was also the same number in gravitational contact during this time, but this was not directed from the electron to the proton, but was directed to all particles both in the near range and to those at a great distance. The decisive point, however, is that it is never possible to measure individual contact in either of these two interactions. Quantum mechanics only records the average sum of the contacts. It can also measure the mean values of inertia, but this is interpreted differently. In the case of quantum mechanical states, we are only dealing with the electrical mean values of these one-sided connections. If we measure them, then we determine the particle for our world. It becomes clear for the moment. If we do not measure it, then it is also deterministically connected to the world, only the networks are extremely large and very volatile and not accessible to our world. Unless we succeed in tapping into the world of the quantum in the quantum computer. But for that we would not be allowed to ask questions from our world on the computer. Not something as solid as a prime number decomposition, but then it would have to be used open-ended. But we'll get to that.

The entanglement thus does not show us a fundamental indeterminism of matter, but only that there are areas to which we have no access. If we then interpret the state of superposition that our electron and positron has seen from above as indeterminate, i.e., free until it is measured, then all this has more to do with the Copenhagen interpretation than with reality. Nils Bohr and Werner Heisenberg have made a break in quantum mechanics. Everything that can no longer be measured is not part of our reality and thus becomes indeterminate. Then entangled particles can have a spooky effect at a distance. What is astonishing about this interpretation, however, is the volatile nature that is suddenly allowed here in physics. Here, as a rational physicist, one can imagine that there are contacts between distant particles that are space- and timeless. Why not apply this to every contact at every exchange between particles?

But if we take the risk and transfer the result of the entanglement to the dense communicative exchange of our entire elementary particle world, then completely different possibilities open up for particle networks to be able to influence objects from great distances.

# 13 | Black Holes or not

Let us now turn our attention to the supposed black holes. We have not quite figured out how consciousness or intelligence actually emerges in the networks. But we have nevertheless learned a lot about the fact that particles are much more complex and multi-layered. Yes, that the particles themselves can also carry the germ of life within themselves, but at least in the large connections they must be much more than just mathematical models. The size of such networked gigantic collections of particles is so indescribably enormous that it is very difficult or perhaps impossible to describe what really happens in these networked areas of matter. There is an electrical connection that creates a lot of stability and there is this volatile gravitational possibility to exchange and store information. Everything we see on earth, but also much of what originated in space on the suns, indicates that matter has actually developed into complex structures over long periods of time. In the suns, nuclear fusion does not just happen, it is consciously desired. In neutron stars of burned-out suns, not much happens to matter anymore, but we have great doubts whether thinking, once it has been created, simply disappears again. A neutron star is certainly no longer physically active, but the networks, if they really do run via gravity and inertia, are not affected. They can also think further as neutrons. It would be different if the matter disappeared in a black hole behind the event horizon. Then not only would no longer be able to think, but also all stored information would be erased. Knowledge and experience would be lost and the networks would be irrevocably dissolved. We have great doubts whether something like this can exist in our universe. Furthermore, we could not reverse this universe and our models would not work anymore. How the original order should be restored in case of a time reversal, if the particles have forgotten everything. Actually, even without thinking networks it would be very strange that matter should be able to leave the web of the world so easily. Preservation of energy and impulses not only means that the balance must be right, but also the arrangement in space, the distribution of energy to near and far particles is also determined. In the order lies after all an energy value, an information, which is equalized in the black hole.

General relativity plays the decisive role in describing how black holes are formed. But it only describes the processes globally. Matter is not treated quantum-mechanically in it and the sum of the particles is summarised as their mass centre. The masses change space, a space that can be stretched and is infinitely fine. This is all very unspecific. It may be that, similar to what is expected in quantum physics, space as the sum of the cross-linked particles acquires a kind of abstract substance for which the term field can be used, but if one examine boundary conditions such as black holes more closely, one also reaches the limits of theory. For example, gravitational collapse only occurs if there is no more counterforce to gravity at the end. For planets and smaller masses, the normal electrically repulsive charges are sufficient to maintain an equilibrium, a long-term stability. The first important counterforce for very large mass concentrations, for example those of solar size, is generated by the pressure of the energy released during nuclear fusion. When the fuel is exhausted and the matter continues to compress after it has expanded or after a supernova, the matter is stopped by the Pauli condition. According to this, no two states may be assumed simultaneously. First an electron gas pressure is created, but the mass continues to increase then the electrons are pressed into the protons and we get a neutron star. The Fermi pressure in neutron stars then represents the last known counterforce according to theory. The neutrons are close together and occupy every minimal possible state in phase space. The density is incredibly high.

After that the physicists do not know any process which causes a counterforce, so they see in the consequence, if further outside mass is added, the matter is getting more and more dense until it comes to a gravitational collapse. At that moment, matter is cut off from the rest of the world. The physicists' gaze is now directed only at the particles deep inside the enormous concentrations of matter. If we assume, however, that matter is not quantum-mechanical particles but rather cross-linked particle systems and that such cross-linking cannot simply be switched off, then we do indeed receive a counterforce to gravity much earlier. Gravity itself arises from the networks. Why should particles want to decompose themselves? It is much more probable that matter that has calmed down in a neutron star no longer seeks its connections among themselves as it becomes denser, but instead moves outwards more and more. And this does not change when more and more matter is added. The more the concentration of particles increases in a region, the more the view wanders outwards, to where they all come from. No particle has forgotten its origin and the further away, or the more foreign particles are around a particle, the more often the contacts are not in the near field but in the distance. This weakens the gravity enormously. We get an increasingly stronger gravitational current to distant matter and a weakening of the connections between them, so that they have a minimum of freedom again. Matter particles are not suicides. The gravitational current does not flow into empty space, it seeks the connections to other accumulations of matter. This increases the gravitational forces within a galaxy. They can therefore turn faster without the suns being ejected. One could even go one step further: These particles in neutron stars, or other clusters of matter, look outward because inside there is too little going on. Nothing exciting happens anymore, so the thinking networks have to look for their variety somewhere else.

If we transfer this to the centres of galaxies, where a supermassive black hole is suspected in the centre of each individual galaxy, then we must also interpret this area in our own way. Today, massive black holes at the centers of galaxies are attributed great importance for the stability of the galaxy and for the development of life. Even if the arguments about their influence are only very vague and indirect, some researchers are convinced that they play their role in evolution. And we also think that when evaluating what is good, what must be sorted out, there must always be some thinking in the background somewhere. In this way, the supposedly supermassive black holes create a structure and a superior support for the objects within. The galaxies as a whole can thus reach very high speeds, and thus move far away from their origin. More and more temporally different matter from different areas of the universe can mix, which increases the diversity. Through the foreign matter, movement arises again.

Nevertheless, we are stuck with our considerations. We, but also the Standard Model, fail because of the almost unbelievably large numbers of particles involved. In the latest research, objects have been found that are supposed to have 80 billion solar masses, although a theoretical limit has actually been calculated for 60 billion solar masses. Astronomers have calculated the masses of black holes from the measurements, which are said to account for 13 billion solar masses at a distance of less than a billion years after the Big Bang. There are galaxies with one, with two and, more recently, with three large black holes at their centres. A few details such as mass size, distance, luminosity or redshift capture something that is actually so enormous that it should be completely incomprehensible to us. So many individual particles, all of which are connected by gravity and are thereby in a causal relationship. No particle is overlooked, none is confused or forgotten in its conservation of energy and momentum. Although the many meets in such a small space, they hold together in an orderly fashion. They manage to create something for themselves and for the whole, so that the wholeness of gravity forms an event horizon. The event horizon is the sum of billions of particles that all together change a space or a field or whatever in such a way that a breakthrough opens. All these incomprehensible quantities of quantum-mechanical single objects create an order that forces space to burst open at one point.

Describing a grain of wheat by taste, weight, moisture or size is easy. We capture the wholeness with a numerical value and a unit. If we are to imagine this interplay at the level of cells or molecules, then we only understand the individual molecule or the structure of a single cell. But how the interaction of the many can function remains a mystery. And it is equally incomprehensible that so many individual particles in such a small area of space all meet, work together and do not lose each other. In fact, we assume that they do so absolutely. It may not know Einstein's equations, but it behaves 100% exactly like this. That's what we all believe, that's what physicists tell us. The problem is not the mathematical equations, that they are only approximations, the problem is only the measurements. The limit of measurability gives the particles their freedom, but the physical formulas apply absolutely to them. So even in the standard idea of supermassive black holes, we are mentally reaching our limits. Not only with the idea that these should also be all thinking objects. The claim that we have one or many consciousnesses of gigantic dimensions at the centre of the Milky Way is clearly incomprehensible, but even if there is no consciousness there, the area there remains incomprehensible. And this is crucial. We can grasp the wholeness with a few numbers and we can investigate the smallest and largest basic size. What lies in between, how the interaction of the many then takes place, what can come out of it, is beyond our imagination. Can something so great, with the changed conditions of the smallest objects in it, then be so much more? Can it move matter, create connections, change the world? Or does it nevertheless remain mechanistic, do we put far too much into it? We will probably not be able to prove it if there is much more to it, but the opposite cannot be proven either. Even black holes in the classical sense are not understood in detail, which is why researchers cannot disprove that there is more to the whole than the details suggest. Perhaps an emergence that has something to think about emerges from the interaction of matter, but if we have no feeling for how an interaction of particles can develop, then we cannot really imagine it. And then scientists prefer to stick to the sober but stable reality of mechanistically running particles.

# 14 something million light years away

The thesis that supermassive black holes are not really black holes, but merely mass accumulations that come close to the Schwarzschild radius but do not reach it, is very presumptuous in view of the many reports and sightings of ever newer and more specific black holes. After all, the first photo of such a monster was even published on 10 April 2019. This photo went through the entire world press and is regarded as the striking proof of the existence of black holes. For the photo, eight radio telescopes on four continents were connected and directed at an object in the Virgo cluster M87, 55 million light years away, which has about 6 billion solar masses, according to estimates from the star motions around it. Estimates based on the motion of the surrounding gas only led to 3 billion solar masses. We can see a bright blurred ring in the photo, which shines brighter in the south than in the north, which is due to a rotation. This luminous ring is either the accretion disk or parts of the jet stream, relativistically consumed. In the dark inner area is then the black hole. This black inner area is 2.5 times, or 5 times at 3 billion solar masses, as large as the event horizon. The event horizon itself is therefore not mapped. What one can see there, however, fits too well with the calculations and simulations of galactic black holes. And this is not only unfavorable for our world view, but also the physicists and astronomers themselves are not entirely happy about it. At the transition to the Schwarzschild radius, quantum mechanics meets the theory of relativity, and since both theories cannot be correct, researchers are literally looking for a discrepancy. It is suspected that the theory of relativity must be changed somewhere, but the analysis of the recording did not reveal anything conspicuous. The photo seems to confirm Einstein's theory once again.

In our world view we would also be pleased about a peculiarity, some kind of conspicuousness, although this photo of an object at a distance of 55 billion light years is not so well resolved that the transition at the event horizon can be seen in detail. The mass concentration does not have to take up much more space than the Schwarzschild radius has, although a much larger area would have made the whole thing clearer. For example, the mass of 3 to 6 billion suns is located in a range within 2.5 to 5 Schwarzschild radius lengths. So, everything is still open in our approach. Matter may have disappeared within the Schwarzschild radius in a singularity, but it may also be distributed in its own ordering structure in a region that is 2.5 or 5 times larger and still no longer shines, because it is only neutrons and these take on every possible state in a limited phase space, but have too few freedoms. After all, neutrons lose their electrical charge and photons can no longer be produced. We then only have the weakened gravity, which is ordered to a maximum degree. Also, the freedom of movement of neutrons is more and more restricted and must be tuned. In order to be able to act at all with so many particles, the view must be directed increasingly outwards. This gives the particles some freedom of movement again. A galactic black hole is not comparable with a stellar black hole. The density of stellar black holes is fantastically high. There is so much mass in so little space that an object that comes too close to it would rip apart simply because of the tidal forces. The side facing the black hole experiences stronger gravity than the side facing away from it, and this difference is so great in the vicinity of black holes that objects inside are crushed. Galactic black holes, if they are very massive, are correspondingly large. The density would then not be so high, rather in the range of air, and the acceleration would be more in the range of the acceleration due to gravity and would thus not tear anything apart. Galactic mass concentrations at the center would thus be quite gentle and could take up space more like a lukewarm wind, if the Schwarzschild radius is not exceeded. If an opposing force is found then the conditions in this dark area are anything but hellish. They would be borderline quantum mechanical, because space is as dense in its states as in a semiconductor, the electrons can only move in bands, we would have the calm of the states there, but the matter would not disappear. On the contrary, it has calmed down again and is captured by an enormous order. A ballet of particles, a dance of neutrons. The flood of particles is still exchanging, but not so much towards the centre, but rather towards the outer regions, or even to atoms that are completely distant. We do not see this connection because we cannot see the gravitation and there are no more glowing free electrons. But if we could make gravity visible, we would see an insane stream of connections to all objects in the galaxy. Perhaps we would then even find that the gravitational current does not go off completely uniformly. It doesn't flow into the void, and perhaps it flows much more to planets with life than to trivial objects like the Moon or Mars. We have not given up yet that there could or should be more in these information streams than we, as sober thinking beings, have conceded to matter. If, in addition to our complexity, there is a mirror image or an associated network at the centre of the galaxies that helps to shape us, then the paths are not laid out statistically, but rather the connecting strands change as interest in us grows, as is the case with our brain. But then we first have to explain how something at a distance of 27,000 light-years can have an influence at all, even without warp drive.

# 15 Live stream between the particle

The distances, the way to other stars is long. From our point of view, everything is at least four light years away, most of it much further, even if we belong to the same system. Other galaxies are much more distant and therefore inaccessible to us. The only thing we experience from there is electromagnetic radiation, some fast particles and if we are lucky, gravitational waves from really big space-time events. We can observe space, but everything we see is already old because it had to travel so long and what we see then was created according to our theory after us, but it is younger according to the big bang theory. But the possibility to have influence is an illusion. If we wanted to send signals to other stars, then it takes decades or even centuries before the signal reaches the other star or a possible answer comes back. How is an exchange to take place, how is influence to be exerted. If we wanted to travel there, the whole thing looks even more gloomy, then the number of years before we arrive there is multiplied by a potency factor. Well, we could put our seeds, our germs from the earth into a satellite, send it self-sufficiently on the search and hope that it will accidentally meet the preform of an earth in a hundred thousand years and then influence life there. Then we would have an influence. But would germs really be able to survive for a hundred thousand years and develop on a dead biotope in an alien world without fertile soil, without the familiar biome from earth? Probably that is impossible. If it were, then perhaps a highly developed species could use technology to analyse the genetic code and use its own means to create an Earth-like being - hypothetically. But an influence on other worlds directly in the present time would be impossible. Large bodies of matter, but also particles or only radiation is much too slow for this. The fastest transmission path simply takes too long for the given distances.

But what does the world look like from the point of view of radiation, electromagnetic and gravitational interaction? Here we have always argued that the contact happens without space and time. That for the duration of $10^{-18}$ the particles lie on top of each other and read out the memory and that no matter how far away the connection is. This is exactly what we still think, but from the point of view of a quantum, the graviton or a contact between two interacting particles, the world cannot be equated so one-to-one with our world. We have said that if something exchanges at the speed of light, and this happens in most processes, then from the point of view of the quantum the connection is timeless, the space in between is not traversed. The quantum knows at the beginning where it arrives, it knows the end before it arrives. But we have no access to such future knowledge without further ado. The quantum knows it, we do not. We have to wait for the time, with us the time in between passes and therefore we artificially put a way in for the quantum, create a virtual space that is real for us. We beings of slowness in endless space transfer our world to the networks of quanta and particles and that doesn't work. Just as thinking in computers is comparatively slow, or even very slow in our brain by comparison, but in a quantum computer it happens almost immediately, we imagine the difference between the thinking of particle networks and ourselves. If there is the networking of matter and in it a thinking or somewhere a second consciousness appears, then it is enormously fast, easy and almost timeless. Then the particles themselves fire. These networks do not work through the movement of the particles or through large macroscopic nerve cells, but here we are in the most elementary, the smallest units that have come together and found ways to complexity. For example, if such a spirit is sitting in a supposed black hole 27,000 light years away, the worldly signals will last 27,000 years, but the contact with the particles on earth is immediate. Particle networks also recognise the end at the beginning, because the particles are superimposed on each other for a short time, again and again. That means there is not only one world view. There is our sluggish world in which everything we see is delayed, the further away the later the images and information arrive, and then there is the world of networks, of connections between particles, and it is always immediate. It might be completely statistical, if there is no second thinking, but then there would be only one evenly distributed gas in our universe. And, it is much more directional and ordered in streams when the particles have discovered consciousness and thinking. Then the objects in the supposed black holes always get to see two pictures. One from the present time in the livestream and one 27,000 years later. The later event is then probably rather uninteresting and more of a statistical nature, but the livestream could mean everything for the intelligences there. They not only take part in our life, but also have the experience. They track our worldliness, our feelings, our knowledge, our love and lust. But also, those of the earth and the other many living beings. We send and receive with our highly complex particles, but not simply into space, but to a second consciousness at a great distance in the centre of the Milky Way. If that is so, if that were so, then the supermassive black holes in the large and small galaxies are the think tanks, the quantum computers in which life is conceived, which then takes place in a few places in a galaxy. Then the sequence would also be better explained. First, preforms of galaxies were formed, in which initially larger and smaller mass accumulations in the centers formed very early. These supposed black holes are then areas where structured thinking happens, which brings order in the form of stars, planets and moons into the system and influences the fusions in the suns, as well as the movements of the bodies and developments on the planets. It searches and it tries to find the complexity, to hold on to it and to develop it. It is always just an attempt out of an understanding, not an ability to control. Matter certainly does not hang like threads on the consciousness, but it is receptive to it. And these connections and interconnections in it are gigantic. This is what we imagine and this is what we would like to prove.

# 16 Quantum computer

Google's researchers published their quantum computer in advance because parts of it had already been leaked. This is the first time in the world that a computer consisting of 54 quantum bits, 53 of which worked, has succeeded in solving a mathematical problem in 3 minutes that would have taken a classic supercomputer 10,000 years to complete. Although the mathematical task was of a more experimental nature and precisely tailored to the advantages of a quantum computer, and the IBM researchers thought that their computers could have calculated the problem in a few days, this does not detract from the basic success and possibly initial breakthrough. Quantum bits, or qubits for short, use the entanglement of particles, the superposition and the possibility of coupling the bits together like a state function of a long wave.

Each qubit works on its own and yet is linked with all other qubits, quantum mechanically entangled. Thus, they each work in parallel and we have a storage capacity of $2^{53}$ possibilities. Each qubit works in its own spatial dimension, giving us an n-dimensional space of possibilities. This storing or reading out happens almost simultaneously because the quantum-mechanical memories are interlocked with each other. If the result is read out in one place, the choice is immediately clear to all the others. The entanglement of objects with far more than just two particles has been possible for a long time. The researchers are relying on various methods to achieve this.

One possibility involves temperatures close to absolute zero. Special conditions apply here. For certain substances, the resistance jumps abruptly towards zero from a certain temperature, it becomes superconducting. If a current has no resistance, it has no exchange. Entangled quantum mechanical objects thus remain connected to each other for a longer period of time. Nevertheless, this does not work forever and the more particles or qubits are involved, the greater the probability that a contact and thus a measurement will occur somewhere and the connection will break down. Seen in this light, the speed at which a quantum computer calculates is not the most amazing thing for it, because quantum computers have to be fast, they don't have much time. It would be better the longer one can hold an entangled state. Just less than three minutes is already a good number.

The susceptibility to errors caused by unintentional premature contact with the world is also precisely the big problem that sets a limit to all fantasy dreams. In practice, we therefore speak of physical qubits, i.e., the real qubits on the processor, with their high susceptibility to errors. With them, a small error leads to a deviation, which is not still read out as a one, as with classical bits, and thus does not falsify our result, but the smallest deviations lead to a false result. We then have a different superimposed state. Besides the physical qubit, logical qubits are now introduced in quantum error correction. Many physical qubits are combined to one logical qubit, which then has a correspondingly reduced error rate. According to the Threshold Theorem, an error-free qubit can be constructed if the error probability is below a certain limit value. However, the numbers for this are very high. To obtain only one logical error-free qubit, 5,000 to 500,000 physical qubits are needed, depending on the assumption and the technique. Maybe the number can be reduced, there is some work to be done, but currently these are the bandwidths that are given. So, for 100 logical qubits we would need 500,000 to 50,000,000 physical qubits.

There is some speculation that these qubits will not be achieved for at least 24 years, but it is not yet certain that it will ever be possible to go beyond the experimental stage. Whether it will ever be more than something of academic interest. Another question is whether we really need full error correction. There are also fields of application where the errors are not as strong or play such a big role. For example, to simulate molecules with a quantum computer. So especially in the field of simulating living things, the errors do not have such a high priority.

Research in Europe is leading the way in the development of qubits with ions. Here the qubits do not consist of chips similar to those used in classical computers: instead of silicon, aluminium and niobium are used. What is important here are the low temperatures. In qubits from ion traps, individual ions are held in electrical fields and then controlled and manipulated by lasers. The whole thing happens in a high vacuum. The type of technology here comes more from the field of atomic clocks, i.e., a technology with extremely high precision. This means that the error rates are the lesser problem. On the one hand, a single ion is a unique quantum object and so isolated it can be controlled well. The difficulty then lies more in building really large registers. Up to now, long chains of ions have worked very well, but for numbers well over 50, one would have liked to arrange the ions in a square pattern. Ionic qubits are less error-prone, but the clock frequency is much slower than that of superconducting qubits. But as long as we don't know exactly what we quantum computers are used for and whether there are also different possible fields where one or the other property is advantageous, it is good that various developments are being continued.

# 17 Quantum computers and networks of life

The first simple quantum computers show that it is indeed possible in principle to use the elementary particles themselves as binary building blocks and to make sensible use of the special nature of quantum mechanical properties. How far we can still improve the technology and the software, how far we can still drive the logical circuits, the networking and the speed in nature is open. So far, however, it has been shown that there is indeed always a way to realize the complicated, the complex and the ever larger. Our search for the ever higher and higher does not stop and somehow, we seem to succeed in getting things going. A little bit we ask ourselves whether the external networks, these other intelligences, don't have their fingers in the pie. The researchers develop something from their experience, their knowledge and skills, then connect it up, let the current flow and suddenly exactly what they had imagined happens and yet they no longer understand why it works at some point after much work and effort. In the end, what really works so smoothly in the machine remains a mystery. And this puzzle not only affects quantum computers, but also computers and many machines, but also networked life in cities, our monetary system, wars or politics. In the end, there is always a successful interaction that changes the world as a whole and that does not lead downhill from an extremely high level, but rather becomes more complicated and larger - there is almost something magical about it all. When we write a book and distribute it en masse, then we have written down our thoughts, recorded them in the time. When it is read by other people, then they understand the content. Then they can do something with it. Devour it or reject it completely, but something in them reacts to the story, to the written abstract word. Some authors are so successful in this that their books become bestsellers and people read them by the millions, even though only one of them developed them and all books say the same thing. It works and nobody knows why. Well, if somebody can't write, then he won't write a bestseller, but not everybody who can write is successful and those who are successful are not necessarily the best. Machines work, computers work and quantum computers may work stable and well at some point, but we don't know why. We feel our thoughts, our ideas, the fleeting stories we make up but can't explain how they come about. And if there is even more behind all the wonderful and mysterious, then we are not able to prove this or the corresponding thoughts in a nerve cell. Then our brains are perhaps more like receivers and transmitters of signals and not the source of them. If everything is so well and stable when it comes to realization, and even develops further and further, then this is probably not least because there is an interest from the other consciousness in going higher and higher. We don't know what it wants, how it works, but we can see how things develop and what changes take place. Computers seem to belong to it and quantum computers probably do too. We also suspect that if a quantum computer represents the deepest possible interface to our world, then there are much more, much more important things hidden in it than we realize. That perhaps quantum computers can be used to tap into the world in the background, even if only in fragments. If we hear about quantum computers, our concern is about encryption, about the fact that the transmission paths are no longer secret and secure. We also learn that quantum computers are very alien and therefore there are not yet as many applications as there are for computers. In everyday life we will therefore probably continue to use our normal processors. But perhaps the most important and uncanny area of application for quantum computers is not the complicated tasks in mathematics or the large databases that are captured at an unprecedented speed, but the possibility of tapping into the networks themselves. Perhaps, quite possibly, quantum computers will one day make it possible not only to detect the networks but even to make contact with them. Maybe we will find a language that both sides can understand, and maybe we will experience completely different new connections or if and where there are other worlds with life. To do this, however, we would have to be open to the fact that there are other abstract intelligences in the networks of matter and we would then have to ask the questions to the quantum computer in a different way. Not how quickly a large number can be broken down into its prime factors, but what the networks have to tell us. We should precisely not get too concrete and should be open to the random. Then it will show if we can understand a completely different foreign thinking. But perhaps it will also prove that this different way of thinking is not so strange. That it is already more familiar and closer to us and we only have to open ourselves to it. Nevertheless, it is something completely different to get in contact with another mind via a quantum computer than via an intuitive feeling.

But no matter what and how it will happen, the future will always remain exciting and thrilling!

# 18 Reference/Index

Fundamentals and calculations for the universe structure as used in the books:

https://christian-hermenau.com/

www.ingramcontent.com/pod-product-compliance
Lightning Source LLC
Chambersburg PA
CBHW071420210526
45465CB00001B/467